YOUR KNOWLEDGE HAS VALUE

AF130047

- We will publish your bachelor's and master's thesis, essays and papers

- Your own eBook and book - sold worldwide in all relevant shops

- Earn money with each sale

Upload your text at www.GRIN.com and publish for free

Bibliographic information published by the German National Library:

The German National Library lists this publication in the National Bibliography; detailed bibliographic data are available on the Internet at http://dnb.dnb.de .

This book is copyright material and must not be copied, reproduced, transferred, distributed, leased, licensed or publicly performed or used in any way except as specifically permitted in writing by the publishers, as allowed under the terms and conditions under which it was purchased or as strictly permitted by applicable copyright law. Any unauthorized distribution or use of this text may be a direct infringement of the author s and publisher s rights and those responsible may be liable in law accordingly.

Imprint:

Copyright © 2017 GRIN Verlag, Open Publishing GmbH
Print and binding: Books on Demand GmbH, Norderstedt Germany
ISBN: 9783668515222

This book at GRIN:

http://www.grin.com/en/e-book/374320/index-theory-of-numbers

Alexander Mircescu

Index Theory of Numbers

GRIN Publishing

GRIN - Your knowledge has value

Since its foundation in 1998, GRIN has specialized in publishing academic texts by students, college teachers and other academics as e-book and printed book. The website www.grin.com is an ideal platform for presenting term papers, final papers, scientific essays, dissertations and specialist books.

Visit us on the internet:

http://www.grin.com/

http://www.facebook.com/grincom

http://www.twitter.com/grin_com

Index Theory of Numbers

Dr. Alexander Mircescu, Munich, Germany

I. Basic Structure

0. Sets

"Under a set we shall understand an aggregation M of distinguishable elements m which are merged to a whole. (Georg Cantor, 1895)."

Hence, we have:

$$M = \{m_a, m_b \cdots\}$$

The elements of M are distinguishable but not ordered.

Two sets are equal if they have the same elements [1].

The building of a set proceeds in two steps. First, we build a multeity $m_a, m_b \cdots$, an accumulation, a range. Then, this multeity is merged to a unity, to a set M [1].

Following definitions are useful.

$$\emptyset \text{ is the empty set}: M = \{\emptyset\}$$

The empty set does not comprise any element.

$$\mathcal{P} \text{ is the power set}: \mathcal{P}(M) = \{a \mid a \subseteq M\}$$

The power set is the set of all subsets of M. If the set M has n elements, then the power set has 2^n elements.

II. Number Systems

1 The Natural Numbers: The Set $\mathbb{N}$

The natural numbers are formed from the set M. The idea is that all elements of the set M shall have an ordering, giving each element of M a "name"/a "number". The natural numbers define a set $\mathbb{N}$ in which one element $0 \in \mathbb{N}$ is distinguished (the first element); and a self-mapping $S: \mathbb{N} \to \mathbb{N}$ (*Successor Function*) exists, such that the following axioms apply [2]:

$(S1)$ S is injective
$(S2)$ $0 \in S(\mathbb{N})$
$(S3)$ If a subset $M \subseteq \mathbb{N}$ exists which contains the 0 and which is self $-$ mapped by S, then $M = \mathbb{N}$

The set $\mathbb{N}$ starts with the point 0 and proceeds with the successor, with its successor, and so forth. We can therefore characterize the natural numbers with an index: *Counter*: $I_\mathbb{N}$:

$$\textbf{Counter}: I_\mathbb{N} = S(m + 1) - S(m)$$
$$with \ m \in \mathbb{N}$$
$$and \ S \ the \ Successor \ Function$$

This index exists for each natural number, since the difference between a successor and its predecessor is always 1. But this index does not exist for the elements of the set M, since said elements are not ordered, such that no subtraction can be carried out. Hence, this index is a characteristic of the natural numbers.

2 The Integers: The Set $\mathbb{Z}$

The natural numbers allow an addition in every case, but do not allow the inverse operation, the subtraction, in any case. In order to remove this disadvantage, the natural numbers are extended to the integers. Every integer can be defined as the difference $(a - b)$ of two natural numbers a, b [2]. The integer $(a - b)$ shall be written as a pair (a, b) [2]. We consider the following relation on $\mathbb{N} \times \mathbb{N}$: $(a, b) \sim (c, d)$ if $a + d = b + c$ (if $(a - b = c - d)$ [2].

The integers are defined as equivalence classes of the relation $\sim$ [2]. The integers set up a commutative group with respect to the addition [2]. The integers set up an integrity ring (commutative, zero division free ring with one-element) [2]. Every integer (a, b) possesses an inverse integer (b, a) [2]. The integers define a set $\mathbb{Z}$ [2]. We can therefore characterize the integers with an index: $\boldsymbol{Orientator}$: $\boldsymbol{I_{\mathbb{Z}}}$:

$$\boldsymbol{Orientator}: \boldsymbol{I_{\mathbb{Z}}} = (a - b) - (b - a)$$
$$with\ a, b \in \mathbb{Z}$$

This index exists for each integer $(a - b)$, since we can always build an inverse integer $(b - a)$. But this index does not exist for the elements of the set $\mathbb{N}$, since said elements do not allow negative numbers, such that the index cannot be always computed. Hence, this index is a characteristic of the integers.

3 The Rational Numbers: The Set $\mathbb{Q}$

The integers allow an addition, a subtraction, and a multiplication in every case, but do not allow a division in any case. In order to remove this disadvantage, the integers are extended to the rational numbers. Every rational number can be described as a quotient $\frac{a}{b}$ of two integers a, b [2]. The rational number $\frac{a}{b}$ shall be written as a pair (a, b) [2]. We consider the following relation on $\mathbb{Z} \times \mathbb{Z}\backslash\{0\}$: $(a, b) \sim (c, d)$ if $a \cdot d = b \cdot c$ $(if$ $\frac{a}{b} = \frac{c}{d})$ [2].

The rational numbers are defined as equivalence classes of the relation $\sim$ [2]. By using the relation $l: \mathbb{Z} \to \mathbb{Q}$, $a \to \frac{a}{1}$ $l(\mathbb{Z}) \subset \mathbb{Q}$ the set $\mathbb{Z}$ is mapped in an isomorph manner to the subring $l(\mathbb{Z}) \subset \mathbb{Q}$ [2]. The field $\mathbb{Q}$ is the smallest field containing $\mathbb{Z}$ as a subring [2]. Every rational number (a, b) possesses an inverse rational number (b, a) [2]. We can therefore characterize the rational numbers with an index: $Interleaver: I_{\mathbb{Q}}$:

$$Interleaver: I_{\mathbb{Q}} = \frac{a}{b} - \frac{b}{a}$$
$$with\ a, b \in \mathbb{Q}$$

This index exists for each rational number $\frac{a}{b}$, since we can always build an inverse rational number $\frac{b}{a}$. But this index does not exist for the elements of the set $\mathbb{Z}$, since said elements do not allow floating point numbers, such that the index cannot be always computed. Hence, this index is a characteristic of the rational numbers.

4 The Real Numbers: The Set $\mathbb{R}$

The rational numbers allow an addition, a subtraction, a multiplication, a division, and an exponentiation in every case, but do not allow the root extraction and the extraction of logarithms in any case. Also transcendental numbers like π or e cannot be defined by rational numbers. In order to remove these disadvantages, the rational numbers are extended to the real numbers. This extension is performed by Dedekind cuts [2].

A Dedekind cut is [1]: Shall $\langle M, < \rangle$ define a linear ordering. A Dedekind cut in $\langle M, < \rangle$ is a pair $\langle L, R \rangle$ with $L, R \subseteq M$ with the following properties:

$(i) L, R \neq \emptyset, L \cap R = \emptyset, L \cup R = M$
(ii) for all $x \in L, y \in R$ we have $x < y$
(iii) $Sup(L) \in L$ if $Sup(L)$ exists

The supremum $Sup(L)$ exists for each rational number. The supremum $Sup(L)$ does not exist for irrational numbers. Here, we have a gap which is closed by the Dedekind cut. Otherwise stated, the Dedekind cuts close all gaps of $\mathbb{Q}$ and lead to the set $\mathbb{R}$ which is complete [2]. We can define the set $\mathbb{R}$ in the following manner [1]: Every non-empty upper bound subset P of $\mathbb{R}$ possesses a supremum (smallest upper bound); there is hence a $s^* \in \mathbb{R}$ with:

(1) $P \leq s^*$
(2) If $s \in \mathbb{R}$ and $P \leq s$, then $s^* \leq s$

s^* is called the supremum of P. We have therefore $s^* = Sup(P)$. We can therefore characterize the real numbers with an index: **Complementator**: $\boldsymbol{I_{\mathbb{R}}}$:

$$\boldsymbol{Complementator}: \boldsymbol{I_{\mathbb{R}}} = Sup(P)$$
$$with\ P \in \mathbb{R},$$
$$P\ a\ non - empty\ upper\ bound\ subset\ of\ \mathbb{R},$$
$$and\ Sup\ the\ Supremum\ (smallest\ upper\ bound)$$

This index exists for each real number, since we can always define a supremum with a Dedekind cut. But this index does not exist for the elements of the set $\mathbb{Q}$, since said elements do not always possess a supremum, such that the index cannot be always computed. Hence, this index is a characteristic of the real numbers.

5 The Complex Numbers: The Set $\mathbb{C}$

The real numbers allow an addition, a subtraction, a multiplication, a division, an exponentiation, a root extraction and an extraction of logarithms in any case, but do not always lead to a solution of algebraic equations. Indeed, when said solution demands the extraction of a square root from a negative real number, then the real numbers do not allow a solution. In order to remove this disadvantage, the real numbers are extended to the complex numbers. This extension is performed in the following manner [2].

The set $\mathbb{R} \times \mathbb{R}$ of all ordered real number pairs $z := (x, y)$ defines by the natural addition $(x_1, y_1) + (x_2, y_2) := (x_1 + x_2, y_1 + y_2)$ an Abel group. We define a multiplication in $\mathbb{R} \times \mathbb{R}$ in the following manner [1]:

$$(x_1, y_1) \cdot (x_2, y_2) := (x_1 \cdot x_2 - y_1 \cdot y_2, x_1 \cdot y_2 + y_1 \cdot x_2)$$

$e := (1,0)$ is the one-element. $i := (0,1) \in \mathbb{C}$ is the imaginary unit with $i^2 = -1$ [2]. We have then:

$$(x_1 + i \cdot y_1) \cdot (x_2 + i \cdot y_2) := ((x_1 \cdot x_2 - y_1 \cdot y_2) + i \cdot (x_1 \cdot y_2 + y_1 \cdot x_2))$$

$e = e_0$ is called the essential real unit. $i = e_1$ is called the essential complex unit.

The set $\mathbb{C}$ is not ordered. We can define [2] a conjugation $\mathbb{C} \times \mathbb{C}$, $z \to z^*$ with $z = x + i \cdot y$ and with $z^* = x - i \cdot y$ as an automorphism which maps $\mathbb{R}$ in itself, and which maps i into the second, in principal equal, zero point $-i$ of $x^2 + 1 = 0$. We can therefore characterize the complex numbers with an index: $\boldsymbol{Conjugator}$: $\boldsymbol{I_{\mathbb{C}}}$:

$$\boldsymbol{Conjugator}: I_{\mathbb{C}} = (a, b) = a - a^*$$
$$with\ a \in \mathbb{C}$$
$$a^*\ the\ complex\ conjugate\ of\ a$$

This index exists for each complex number including the essential complex unit, since we can always build a complex conjugate for any such complex number. But this index does not exist for the elements of the set $\mathbb{R}$, since said elements do not possess an imaginary unit, such that the index cannot be computed. Hence, this index is a characteristic of the complex numbers.

6 The Quaternions: The Set $\mathbb{H}$

The complex numbers allow the solution of any algebraic equation by introducing one imaginary unit. The question arises, if one can generalise the complex numbers by introducing further imaginary units. The first generalisation leads to the quaternions [2]. We shall define the algebra $\mathbb{H}$ of quaternions as: In the four dimensional $\mathbb{R}$ vector space $\mathbb{R}^4$ the standard basis is defined by the ordered real quadruples: $e_1 = (1,0,0,0); e_2 = (0,1,0,0); e_3 = (0,0,1,0); e_4 = (0,0,0,1)$. e_1 is the one-element. The nine products $e_\mu \cdot e_\nu$ with $\mu, \nu = 1,2,3$ are defined by the following multiplication table of Figure 1.

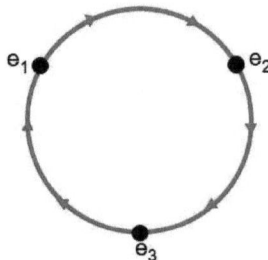

Figure 1: Multiplication Table of Quaternions

We have: $e_1{}^2 = e_2{}^2 = e_3{}^2 = e_1 \cdot e_2 \cdot e_3 = -1$. We also have $e_1 \cdot e_2 = e_3$.

e_2 is called the essential quaternion unit.

The Hamiltonian algebra $\mathbb{H}$ is an associative division algebra [2]. We have $e_1 \cdot e_2 = -e_2 \cdot e_1$ [2]. We can therefore characterize the quaternions with an index: $\textbf{\textit{Commutator}}: I_{\mathbb{H}}$:

$$\textbf{\textit{Commutator}}: I_{\mathbb{H}} = (a, b) = a \cdot b - b \cdot a$$
$$\textit{with } a, b \in \mathbb{H}$$

This index exists and is not zero for each quaternion including at least two complex units, since said quaternions do not commute. But this index is always zero for the elements of the set $\mathbb{C}$, since said elements always commute. Hence, this index is a characteristic of the quaternions.

7 The Octonions: The Set $\mathbb{O}$

The quaternions are the first generalization of the complex numbers. The second generalization leads to the octonions [2]. For elements of $\mathbb{H} \times \mathbb{H}$ we define a product [2]:

$$x, y := (x_1, x_2) \cdot (y_1, y_2) := \left(x_1 \cdot y_1 - y^*_2 \cdot x_2, x_2 \cdot y^*_1 + y_2 x_1\right) \text{with } x, y \epsilon \mathbb{H}$$

The set $\mathbb{O}$ defines an alternative division algebra [2]. The 49 products $e_\mu \cdot e_\nu$ with $\mu, \nu = 1 \cdots 7$ are defined by the following multiplication table of Figure 2.

e_4 is called the essential octonion unit.

We have 7 quaternionic groups in the octonions:

$(e_1, e_2, e_3), (e_1, e_4, e_5), (e_2, e_4, e_6), (e_3, e_4, e_7), (e_5, e_3, e_6), (e_6, e_1, e_7), (e_7, e_2, e_5).$

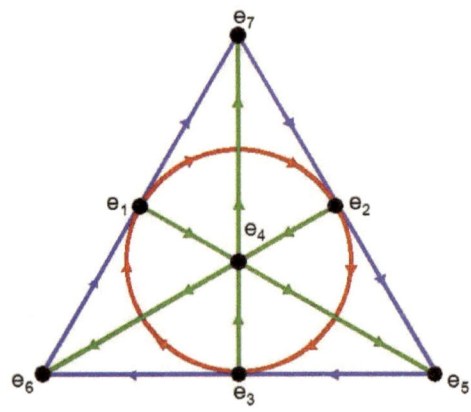

Figure 2: Multiplication Table of Octonions

We have: $\prod_{i=1}^{i=7} e_i = \left(\left(\left(\left((e_1 \cdot e_2) \cdot e_3\right) \cdot e_4\right) \cdot e_5\right) \cdot e_6\right) \cdot e_7 = -1$. The alternative algebra implies [2]:

left alternativity: $a \cdot (a \cdot b) = (a^2) \cdot b$
right alternativity: $a \cdot (b^2) = (a \cdot b) \cdot b$

We have $e_1 \cdot (e_2 \cdot e_4) = -(e_1 \cdot e_2) \cdot e_4$ [2]. We can therefore characterize the octonions with an index: *Associator:* $I_{\mathbb{O}}$:

$$\textbf{\textit{Associator:}} \, \boldsymbol{I_{\mathbb{O}}} = (a, b, c) = a \cdot (b \cdot c) - (a \cdot b) \cdot c$$
$$with \, a, b, c \in \mathbb{O}$$

This index exists and is not zero for each octonion including units from different quaternionic groups, since said octonions do not associate. But this index is always zero for the elements of the set $\mathbb{H}$, since said elements always associate. Hence, this index is a characteristic of the octonions.

The four algebras $\mathbb{R}, \mathbb{C}, \mathbb{H}, and \, \mathbb{O}$ are the only division algebras; hence algebras without zero divisors [2].

8 The Sedenions: The Set $\mathbb{S}$

The generalization of the octonions leads to the sedenions. The set $\mathbb{S}$ of the sedenions is built by the product of $\mathbb{O} \times \mathbb{O}$ by [3]:

$$x, y := (x_1, x_2) \cdot (y_1, y_2) := \left(x_1 \cdot y_1 - y^*_2 \cdot x_2, x_2 \cdot y^*_1 + y_2 x_1\right) \text{with } x, y \epsilon \mathbb{O}$$

The 225 products $e_\mu \cdot e_\nu$ $with$ $\mu, \nu = 1 \cdots 15$ are defined by the following multiplication table depicted in Figure 3 [3].

e_8 is called the essential sedenion unit.

We have octonionic groups in the sedenions.

×	e_0	e_1	e_2	e_3	e_4	e_5	e_6	e_7	e_8	e_9	e_{10}	e_{11}	e_{12}	e_{13}	e_{14}	e_{15}
e_0	e_0	e_1	e_2	e_3	e_4	e_5	e_6	e_7	e_8	e_9	e_{10}	e_{11}	e_{12}	e_{13}	e_{14}	e_{15}
e_1	e_1	$-e_0$	e_3	$-e_2$	e_5	$-e_4$	$-e_7$	e_6	e_9	$-e_8$	$-e_{11}$	e_{10}	$-e_{13}$	e_{12}	e_{15}	$-e_{14}$
e_2	e_2	$-e_3$	$-e_0$	e_1	e_6	e_7	$-e_4$	$-e_5$	e_{10}	e_{11}	$-e_8$	$-e_9$	$-e_{14}$	$-e_{15}$	e_{12}	e_{13}
e_3	e_3	e_2	$-e_1$	$-e_0$	e_7	$-e_6$	e_5	$-e_4$	e_{11}	$-e_{10}$	e_9	$-e_8$	$-e_{15}$	e_{14}	$-e_{13}$	e_{12}
e_4	e_4	$-e_5$	$-e_6$	$-e_7$	$-e_0$	e_1	e_2	e_3	e_{12}	e_{13}	e_{14}	e_{15}	$-e_8$	$-e_9$	$-e_{10}$	$-e_{11}$
e_5	e_5	e_4	$-e_7$	e_6	$-e_1$	$-e_0$	$-e_3$	e_2	e_{13}	$-e_{12}$	e_{15}	$-e_{14}$	e_9	$-e_8$	e_{11}	$-e_{10}$
e_6	e_6	e_7	e_4	$-e_5$	$-e_2$	e_3	$-e_0$	$-e_1$	e_{14}	$-e_{15}$	$-e_{12}$	e_{13}	e_{10}	$-e_{11}$	$-e_8$	e_9
e_7	e_7	$-e_8$	e_5	e_4	$-e_3$	$-e_2$	e_1	$-e_0$	e_{15}	e_{14}	$-e_{13}$	$-e_{12}$	e_{11}	e_{10}	$-e_9$	$-e_8$
e_8	e_8	$-e_9$	$-e_{10}$	$-e_{11}$	$-e_{12}$	$-e_{13}$	$-e_{14}$	$-e_{15}$	$-e_0$	e_1	e_2	e_3	e_4	e_5	e_6	e_7
e_9	e_9	e_8	$-e_{11}$	e_{10}	$-e_{13}$	e_{12}	e_{15}	$-e_{14}$	$-e_1$	$-e_0$	$-e_3$	e_2	$-e_5$	e_4	e_7	$-e_6$
e_{10}	e_{10}	e_{11}	e_8	$-e_9$	$-e_{14}$	$-e_{15}$	e_{12}	e_{13}	$-e_2$	e_3	$-e_0$	$-e_1$	$-e_6$	$-e_7$	e_4	e_5
e_{11}	e_{11}	$-e_{10}$	e_9	e_8	$-e_{15}$	e_{14}	$-e_{13}$	e_{12}	$-e_3$	$-e_2$	e_1	$-e_0$	$-e_7$	e_6	$-e_5$	e_4
e_{12}	e_{12}	e_{13}	e_{14}	e_{15}	e_8	$-e_9$	$-e_{10}$	$-e_{11}$	$-e_4$	e_5	e_6	e_7	$-e_0$	$-e_1$	$-e_2$	$-e_3$
e_{13}	e_{13}	$-e_{12}$	e_{15}	$-e_{14}$	e_9	e_8	e_{11}	$-e_{10}$	$-e_5$	$-e_4$	e_7	$-e_6$	e_1	$-e_0$	e_3	$-e_2$
e_{14}	e_{14}	$-e_{15}$	$-e_{12}$	e_{13}	e_{10}	$-e_{11}$	e_8	e_9	$-e_6$	$-e_7$	$-e_4$	e_5	e_2	$-e_3$	$-e_0$	e_1
e_{15}	e_{15}	e_{14}	$-e_{13}$	$-e_{12}$	e_{11}	e_{10}	$-e_9$	e_8	$-e_7$	e_6	$-e_5$	$-e_4$	e_3	e_2	$-e_1$	$-e_0$

Figure 3: Multiplication Table of Sedenions

The multiplication of sedenions is not commutative, not associative, and not alternative [2]. Said multiplication is only power associative, $a^{i+j} = (a)^i \cdot (a)^j$, and flexible, $(a, b, a) = a \cdot (b \cdot a) - (a \cdot b) \cdot a$ [2]. The multiplication of sedenions possesses zero divisors depicted in Figure 4 [3].

GoTo#1	Based on Octonion Triplet$(1,2,3)$-Automorpheme:$(1,2,3,12,13,14,15)$			
	$(1{+}13)(2{-}14)$	$(1{+}14)(2{+}13)$	$(1{-}12)(2{-}15)$	$(1{-}15)(2{+}12)$
	$(2{-}14)(3{+}15)$	$(2{+}13)(3{-}12)$	$(2{-}15)(3{-}14)$	$(2{+}12)(3{+}13)$
	$(3{+}15)(1{-}13)$	$(3{-}12)(1{-}14)$	$(3{-}14)(1{+}12)$	$(3{+}13)(1{+}15)$
GoTo#2	Based on Octonion Triplet$(1,4,5)$-Automorpheme:$(1,4,5,10,11,14,15)$			
	$(1{+}14)(4{-}11)$	$(1{+}11)(4{+}14)$	$(1{-}15)(4{-}10)$	$(1{-}10)(4{+}15)$
	$(4{-}11)(5{+}10)$	$(4{+}14)(5{-}15)$	$(4{-}10)(5{-}11)$	$(4{+}15)(5{+}14)$
	$(5{+}10)(1{-}14)$	$(5{-}15)(1{-}11)$	$(5{-}11)(1{+}15)$	$(5{+}14)(1{+}10)$
GoTo#3	Based on Octonion Triplet$(1,7,6)$-Automorpheme:$(1,7,6,10,11,12,13)$			
	$(1{+}11)(7{-}13)$	$(1{+}13)(7{+}11)$	$(1{-}10)(7{-}12)$	$(1{-}12)(7{+}10)$
	$(7{-}13)(6{+}12)$	$(7{+}11)(6{-}10)$	$(7{-}12)(6{-}13)$	$(7{+}10)(6{+}11)$
	$(6{+}12)(1{-}11)$	$(6{-}10)(1{-}13)$	$(6{-}13)(1{+}10)$	$(6{+}11)(1{+}12)$
GoTo#4	Based on Octonion Triplet$(2,4,6)$-Automorpheme:$(2,4,6,9,11,13,15)$			
	$(2{+}15)(4{-}9)$	$(2{+}9)(4{+}15)$	$(2{-}13)(4{-}11)$	$(2{-}11)(4{+}13)$
	$(4{-}9)(6{+}11)$	$(4{+}15)(6{-}13)$	$(4{-}11)(6{-}9)$	$(4{+}13)(6{+}15)$
	$(6{+}11)(2{-}15)$	$(6{-}13)(2{-}9)$	$(6{-}9)(2{+}13)$	$(6{+}15)(2{+}11)$
GoTo#5	Based on Octonion Triplet$(2,5,7)$-Automorpheme:$(2,5,7,9,11,12,14)$			
	$(2{+}9)(5{-}14)$	$(2{+}14)(5{+}9)$	$(2{-}11)(5{-}12)$	$(2{-}12)(5{+}11)$
	$(5{-}14)(7{+}12)$	$(5{+}9)(7{-}11)$	$(5{-}12)(7{-}14)$	$(5{+}11)(7{+}9)$
	$(7{+}12)(2{-}9)$	$(7{-}11)(2{-}14)$	$(7{-}14)(2{+}11)$	$(7{+}9)(2{+}12)$
GoTo#6	Based on Octonion Triplet$(3,4,7)$-Automorpheme:$(3,4,7,9,10,13,14)$			
	$(3{+}13)(4{-}10)$	$(3{+}10)(4{+}13)$	$(3{-}14)(4{-}9)$	$(3{-}9)(4{+}14)$
	$(4{-}10)(7{+}9)$	$(4{+}13)(7{-}14)$	$(4{-}9)(7{-}10)$	$(4{+}14)(7{+}13)$
	$(7{+}9)(3{-}13)$	$(7{-}14)(3{-}10)$	$(7{-}10)(3{+}14)$	$(7{+}13)(3{+}9)$
GoTo#7	Based on Octonion Triplet$(3,6,5)$-Automorpheme:$(3,6,5,9,10,12,15)$			
	$(3{+}10)(6{-}15)$	$(3{+}15)(6{+}10)$	$(3{-}9)(6{-}12)$	$(3{-}12)(6{+}9)$
	$(6{-}15)(5{+}12)$	$(6{+}10)(5{-}9)$	$(6{-}12)(5{-}15)$	$(6{+}9)(5{+}10)$
	$(5{+}12)(3{-}10)$	$(5{-}9)(3{-}15)$	$(5{-}15)(3{+}9)$	$(5{+}10)(3{+}12)$

Figure 4: The 84 Zero Divisors of the Sedenions

Sedenions set up a non-commutative Jordan algebra [2]. It is proven that the space of pairs of norm-one sedenions that multiply to zero is homeomorphic to the compact

13

form of the exceptional Lie group G_2 [3]. We can therefore characterize the sedenions with an index: **Alternator**: $I_\mathbb{S}$:

$$Alternator: I_\mathbb{S} = (a, a, b) = a \cdot (a \cdot b) - (a^2) \cdot b$$
$$with\ a, b\ \in \mathbb{S}$$
$$or\ higher\ dimensional\ complex\ numbers$$

This index exists and is not zero for each sedenion including units from different octonionic groups, since said sedenions are not alternative. But this index is always zero for the elements of the set $\mathbb{O}$, since said elements are always alternative. Hence, this index is a characteristic of the sedenions.

Remark 1: If $a \cdot b = 0$ such that $a \cdot (a \cdot b) = 0$, for $a, b \in \mathbb{s}$, hence if $a \cdot b$ is a zero divisor, we have $a^2 \neq 0$, since a^2 is never a zero divisor according to Figure 4. Then, we also have $(a^2) \cdot b \neq 0$, since $(a^2) \cdot b$ is also never a zero divisor according to Figure 4. Hence, the **Alternator**: $I_\mathbb{S}$ is always existing and not zero for every sedenion including units from different octonionic groups, even if $a \cdot b$ is a zero divisor.

If we further generalize the sedenions to higher dimensional complex numbers, no new algebraic structures occur [2]. We can finally define an index **Flexibilator**: $I_\aleph$ which is zero for all numbers:

$$Flexibilator: I_\aleph = (a, b, a) = a \cdot (b \cdot a) - (a \cdot b) \cdot a = 0$$
$$with\ a, b\ \in \mathbb{N}, \mathbb{Z}, \mathbb{Q}, \mathbb{R}, \mathbb{C}, \mathbb{H}, \mathbb{O}, \mathbb{S}$$
$$or\ higher\ dimensional\ complex\ numbers$$

9 Summary of the Structures $\mathbb{N}, \mathbb{Z}, \mathbb{Q}, \mathbb{R}, \mathbb{C}, \mathbb{H}, \mathbb{O}$ and $\mathbb{S}$

The structures of the sets $\mathbb{N}, \mathbb{Z}, \mathbb{Q}, \mathbb{R}, \mathbb{C}, \mathbb{H}, \mathbb{O}$ and $\mathbb{S}$ are summarized in Figure 5 below.

FORM	MEASURE	OPERATOR
$\mathbb{N}$: Open	$\mathbb{Z}$: Discrete (steps)	$\mathbb{R}$: Complementation
$\mathbb{Z}$: Closed	$\mathbb{Q}$: Dense (gaps)	$\mathbb{C}$: Conjugation
	$\mathbb{R}$: Continuous (complete)	$\mathbb{H}$: Commutation
		$\mathbb{O}$: Association
		$\mathbb{S}$: Alternation

Figure 5: The structures of the sets $\mathbb{N}, \mathbb{Z}, \mathbb{Q}, \mathbb{R}, \mathbb{C}, \mathbb{H}, \mathbb{O}$ and $\mathbb{S}$

The form, measure, and operator define three independent features, in the sense that they can be combined independently from one another. Hence, we can use sedenions with rational coefficients which are always positive.

The elements of a single feature cannot be, however, combined. Indeed, we cannot combine an open with a closed form, since they are excluding each other. Once we have selected the form, its index (counter $I_{\mathbb{N}}$, orientator $I_{\mathbb{Z}}$) is decided upon and cannot be changed.

Also, we cannot combine a discrete, a dense, and a continuous measure with one another, since they are also excluding each other. Once we have selected the measure, its index (orientator $I_{\mathbb{Z}}$, interleaver $I_{\mathbb{Q}}$, complementator $I_{\mathbb{R}}$) is decided upon and cannot be changed.

Finally, once we have selected how many dimensions a number has, the operators leading to its index (complementator $I_{\mathbb{R}}$, conjugator $I_{\mathbb{C}}$, commutator $I_{\mathbb{H}}$, associator $I_{\mathbb{O}}$, alternator $I_{\mathbb{S}}$) are decided upon and cannot be changed.

Bibliography

[1]: Deiser: Einführung in die Mengenlehre, 2. Auflage, Springer, 2004.

[2]: Ebbinghaus et al: Zahlen, 3. Auflage, Springer, 1992.

[3]: Raoul E. Cawagas; On the Structure and Zero Divisors of the Cayley-Dickson Sedenion Algebra; Discussiones Mathematicae; General Algebra and Applications 24 (2004) 251-265.

YOUR KNOWLEDGE HAS VALUE

- We will publish your bachelor's and
  master's thesis, essays and papers

- Your own eBook and book -
  sold worldwide in all relevant shops

- Earn money with each sale

Upload your text at www.GRIN.com
and publish for free